Animal Babies

Eaglets

by Meg Gaertner

FOCUS READERS
SCOUT

www.focusreaders.com

Focus Readers is distributed by North Star Editions:
sales@northstareditions.com | 888-417-0195

Produced for Focus Readers by Red Line Editorial.

Photographs ©: Dimchan/Dreamstime, cover, 1; Cecil Bo Dzwowa/Shutterstock Images, 4; Jon C. Beverly/Shutterstock Images, 7, 16 (top right); Rocky Grimes/Shutterstock Images, 9 (top); Brian E. Kushner/Shutterstock Images, 9 (bottom), 16 (bottom left); Tom Tietz/Shutterstock Images, 11, 16 (bottom right); Michael Meshcheryakov/Shutterstock Images, 13; Ingvar Grimsmo/Shutterstock Images, 15; PhotoTalk/iStockphoto, 16 (top left)

Library of Congress Cataloging-in-Publication Data
Names: Gaertner, Meg, author.
Title: Eaglets / by Meg Gaertner.
Description: Lake Elmo, MN : Focus Readers, [2020] | Series: Animal babies | Audience: K to Grade 3. | Includes index.
Identifiers: LCCN 2018056931 (print) | LCCN 2018058976 (ebook) | ISBN 9781641859530 (PDF) | ISBN 9781641858847 (ebook) | ISBN 9781641857468 (hardcover) | ISBN 9781641858151 (paperback)
Subjects: LCSH: Eagles--Infancy--Juvenile literature.
Classification: LCC QL696.F32 (ebook) | LCC QL696.F32 G335 2020 (print) | DDC 598.9/421392--dc23
LC record available at https://lccn.loc.gov/2018056931

Printed in the United States of America
Mankato, MN
May, 2019

About the Author

Meg Gaertner is a children's book editor and author. She lives in Minneapolis, Minnesota. When not writing, she is usually dancing or spending time outside.

Table of Contents

feathers

First Days

Eaglets are baby eagles.

A mother lays eggs.

The eggs hatch.

The eaglets

have **feathers**.

Eaglets stay in the **nest**.

They stay near their mother.

She keeps them warm.

mother
nest

Changes

Eaglets get older.

Their mother feeds them.

They eat **fish**.

mother
fish

Eaglets flap their **wings**.

They grow longer feathers.

They are ready to fly.

wing

Eaglets grow fast.

They fly.

They learn to find food.

Growing Up

Eaglets keep growing.

They leave the nest.

They live alone.

Glossary

feathers

nest

fish

wings

Index